Living on the Edge

Deep Oceans

WENDY PFEFFER

BENCHMARK BOOKS

MARSHALL CAVENDISH
NEW YORK

NEW HANOVER COUNTY
PUBLIC LIBRARY
201 CHESTNUT STREET
WILMINGTON, NC 28401

Contents

Deep Oceans — 5

Hatchet Fish — 14

Deep-Sea Anglerfish — 15

Deep-Sea Squid — 17

Bottom Dwellers — 22

Hot Vents and Cold Seeps — 25

Other Sea Creatures Adapt and Survive — 30

Glossary — 36

Find Out More — 37

Index — 39

In areas on the deep ocean floor where melted rock is exposed, water becomes so hot that it shoots upward. Scientists call such areas hydrothermal vents.

Deep Oceans

Far beneath the surface of the world's oceans conditions are extremely harsh. It's hard to imagine how anything can survive down there. The water pressure is enormous. Under its weight humans would be crushed instantly. It's completely dark. There are poisonous metals and gases. The water is near freezing except where it makes contact with liquid rock beneath the ocean floor. Then it reaches scalding temperatures. Yet, amazingly, the deep ocean is home to thousands of species of sea creatures.

Most sea creatures move through the ocean's upper levels. Along the shore, where the ocean is only a few inches deep, tiny crabs scurry about and disappear under wet sand. Sea stars and certain kinds of jellyfish live in slightly deeper water. Farther down there are sharks, dolphins, and whales. None of these creatures, not even deep-diving whales, ever comes close to reaching the deepest parts of the ocean. Most of them occupy the two upper layers of the ocean: the sunlight zone and the twilight zone.

Crabs live where ocean meets land.

The sunlight zone is the layer of ocean from the surface down to about 650

Plants cannot survive below the sunlight zone.

feet (200 m). This is the only ocean layer in which plants can live. Below this zone there is not enough sunlight to support plant life. Plants growing in the sunlight zone, like plants on land, give off oxygen in their special food-making process, called photosynthesis. Therefore, the sunlight zone has the most oxygen of any ocean layer. It's not surprising that hundreds of species of sea animals, all of which need oxygen to survive, crowd this zone.

Below the sunlight zone is the twilight zone, which stretches from about 650 feet (200 m) to depths of about 3,300 feet (1,000 m). Not enough sunlight filters through to the twilight zone for plants to survive. It is a dimly lit world, much less crowded than the sunlight zone.

Fiddler Crab
(shore)

Crystal Jellyfish
(300 ft/92 m)

Mako Shark
(650–1,300 ft/198–396 m)

Anglerfish
(3,000 ft/914 m)

Giant Squid
(6,000–15,000 ft/1,829–4,572 m)

Blue Whale
(400 ft/122 m)

Hatchet Fish
(600–3,000 ft/183–914 m)

Rattails
(3,000–13,000 ft/914–3,960 m)

Cusk Eel
(20,000–30,000 ft/6,100–9,144 m)

Tubeworms
(13,000–20,000 ft/
3,960–6,100 m)

The twilight may sound pretty deep, but not if you consider how deep the ocean really is. The deepest known ocean trench plunges down about 36,000 feet (11,034 m). The water in this trench contains almost no oxygen and, as you can imagine, is pitch-black. The water pressure alone would instantly kill a human. But thanks to inventions developed in the last fifty years, such as deep-diving submersibles, we are now able to study the ocean's depths.

Over the course of thousands of years, deep-sea creatures have adapted to their cold, dark environment. Deep-sea squid swim toward the surface at night. They feed on tiny plants and animals called plankton and return to the depths at dawn. Some creatures stay deep and feed on marine snow, little bits of dead plants and animals drifting down from above. Other creatures in the least welcoming parts of the deep ocean do not eat at all. They get their food simply by absorbing the sugars that certain bacteria produce.

Almost every creature that dwells in the ocean's dark depths is able to produce light. This ability is called bioluminescence. Flashlight fish have constantly

Flashlight fish have glowing patches under their eyes. They use the light patches to confuse predators and attract mates.

Submersibles can dive to more than 20,000 feet (6,100 m).

A deep-sea shrimp squirts bright glowing ink into the face of a predator.

Squids produce light that fools creatures swimming below. When looking up, the lower fish believe they are seeing moonlight.

glowing patches under their eyes that they can cover and uncover at will. To do this they simply draw a thick membrane, like an eyelid, over their light-producing organs. When a predator approaches, a flashlight fish will swim in one direction and then another, blinking its lights rapidly to confuse its attacker.

Certain deep-sea shrimp escape predators by squirting bioluminescent ink. The bright light of the ink stuns predators, whose eyes are used to the complete darkness of the deep, and leaves them temporarily blind. Before the light fades, the shrimp swim to safety.

Bioluminescence helps deep-sea animals catch prey. The black dragonfish has a luminous strand of flesh dangling from its chin that it uses to attract fish. Fish that are drawn to the dangerous mouth do not escape.

Hatchet Fish

No longer than a finger, hatchet fish live in the twilight zone. This zone stretches from 600 feet (183 m) to about 3,000 feet (914 m). Hatchet fish recognize one another by their special pattern of lights. The hatchet fish's light-producing organ has a reflector and a lens like a flashlight. These focus the beam of light, making it brighter.

A hatchet fish's belly is studded with many light organs. These help camouflage, or disguise, the fish. How? A hatchet fish detects the color and strength of any light that shines on its back and then sends the same color and strength of light from its belly. This makes it almost invisible to creatures swimming below.

Light is the hatchet fish's best defense. The hatchet fish's curved eyes are also a great help, because they can see in most directions. Their sharp eyes and powerful jaws let hatchet fish find and catch the tiny shrimp they feed on.

Hatchet fish live in the twilight zone.

Deep-Sea Anglerfish

Anglerfish attract prey using a lighted lure attached to the top of their heads.

Deep-sea anglerfish can live more than 1 mile (1.6 km) below the ocean's surface. Growing from the top of anglerfish's heads is a long, bony structure with a fleshy, glowing tip. It looks like a fishing pole with dangling bait. Fish are attracted to the glowing tip and try to snatch it. But when a fish comes close to an angler's bait, it's the fish that ends up getting snatched. Female deep-sea anglers are larger than males. A male finds a female by smell and attaches himself to her. After breeding, he stays attached and shares her circulatory system. What a way to survive!

Squid use their arms to catch prey, not to get from place to place.

Deep-Sea Squid

One of the most fascinating creatures of the deep is the squid. A squid does not use fins to pull itself through water. Instead, it sucks in water through its mantle, squeezes it over its gills, and then forces it out of its funnel. This makes the squid's torpedo-shaped body take off like a jet. It can speed through the water in any direction just by rotating its funnel.

Squid eyes are large and glow pale blue. Some 6-inch (15-cm) squid have eyes the same size as a person's. Squid eyes are remarkable. They can look in front and back at the same time and even watch overhead. Their eyes can detect the faintest glimmer of light.

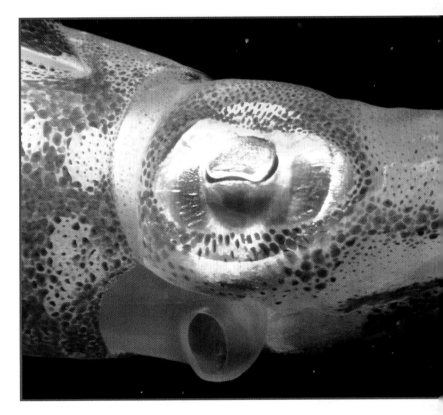

A squid's large eyes can see in more than one direction at a time.

Unlike squid that live closer to the surface, deep-sea squid are adapted to total darkness. Their tentacles can produce bright beams of light. Squid use these bright beams, which are like headlights, to catch prey. After locating a potential victim in the dark and moving close to it, a squid will quickly turn on its light. The bright light stuns the prey and gives the squid plenty of time to attack. If an enemy such as a shark, whale, or seal happens to be swimming nearby, the squid will turn off its lights. Squid also have other defenses against predators. They can change colors to blend in with their surroundings. This allows them to escape easily when they are at shallow depths and can be seen by predators. Squid also change colors to communicate with one another. When they're excited, they flash a deep red. Squid are well equipped to protect their eggs. Females cover their eggs with a bacterial coating that is

Giant Squid

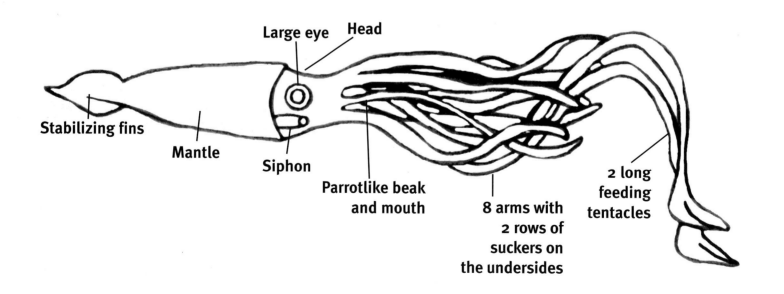

harmful to predators. Sea animals, such as crabs, stay away. But when the young squid hatch, they make tasty treats for many creatures of the deep.

Scientists believe that giant squid, close relatives of deep-sea squid, prowl the ocean more than a mile below the surface. They can grow up to 60 feet (18 m) long and have the largest eyes of any animal. Imagine a creature as long as a tractor-trailer with eyes as large as soccer balls.

No scientist has ever seen a giant squid in its natural habitat. They have only seen squid remains, often in the stomachs of dead sperm whales. Sperm whales can hold their breath for an hour and a half, so they can dive deep in the ocean. Since scientists have never seen a sperm whale catch a giant squid, they don't know how it's done. Maybe whales find squid with echolocation (the use of echoes to find objects), and then stun them with sound.

Recently, a previously unknown squid species was found near the sea floor of four deep-ocean basins. One researcher described these squid as looking

A giant squid netted off the coast of New Zealand. Scientists have never seen a giant squid alive.

like a pair of elephant ears on top of moon landing gear. Up to 23 feet (7 m) long, these slimy creatures have large fins and extremely long arms. They do not seem afraid of submersibles, the small craft that take researchers to the ocean's depths.

Squid communicate with one another by changing colors. When squid are excited, they turn red.

Bottom Dwellers

The ocean bottom in the deep sea is covered with soft, watery ooze, the tiny remains of plants and animals. Because it is difficult to move around in the ooze, bottom dwellers are usually sitters, creepers, burrowers, and floaters. Sea slugs creep along the ocean floor on just one foot. An animal called a sea

Sea pens stand still on the ocean bottom, eating the decaying bits of plants and animals that drift down from above.

pen stands on slender stalks with swollen bulbs anchored in the ooze and feeds on the marine snow drifting down from above.

The ocean bottom is also home to sea cucumbers, sea stars, sponges, starfish, and marine worms. Marine worms inhabit almost every inch of the ocean bottom, feeding on sediment and anything else that's around. Rattails, named for their long, tapered tails, are the most common fish found near the deep ocean bottom. They cruise just above the bottom, looking for animals—dead or alive—to eat.

Flatfish match their color to their background. They hide from predators by lying low and blending in.

The flatfish begins life with one eye on each side of its body. Over the next few weeks, as it lies on its side, one eye moves until both eyes are on the same side. The eyeless side becomes its underside.

The world's deepest known colony of clams lives at 20,000 feet (6,000 m). At this depth the water pressure—the weight of the water—can be about 6 tons per square inch (866 kg per cm^2). A human would be crushed by pressure this great.

Scalding water shoots through chimneylike formations above hot vents.

Hot Vents and Cold Seeps

Hot vents are places in the ocean floor where scalding water spouts out of chimneylike mounds, some towering as high as 180 feet (55 m). Here's how it happens.

Almost two miles (3.3 km) down, ocean water seeps into cracks in the ocean floor. There it comes in contact with the hot melted rock beneath Earth's crust. The water temperature shoots up to 750 degrees Fahrenheit (400° C). The water gets so hot that minerals from nearby rocks dissolve in it. The heated water rises, and scalding, mineral-rich fluids shoot up.

The scalding water mixes with ocean water that is near freezing, and the minerals scatter into the darkness. Over years, the minerals pile up and cement together, making chimneys called "black smokers." Despite the extreme temperatures around hot vents, the water is teeming with 8-inch (20-cm) tubeworms, blindfish, white crabs, spider crabs, rattail fish, and clams as big as hubcaps.

The secret to these creatures' survival is bacteria. The mineral-rich water around hot vents allows bacteria to grow and reproduce. Animals living near hot vents use the sugars produced by bacteria as food, and these animals in turn become food for other sea animals. In the sunlight zone, plant life is the foundation of marine communities. But near hot vents the foundation is bacterial life that can flourish without the slightest bit of sunlight.

Some creatures use the minerals in the water directly. Tubeworms take minerals from the vent waters and squirt it out to form hollow tubes. These tubes become their homes. Hot vents, however, add poisonous metals

Tubeworms make hollow tubes from minerals in the water near hydrothermal vents. The tubes protect the worms from predators.

Rattails swim just above the ocean bottom looking for food.

to the water. Hot vent animals either expel the metals with mucus or depend on metal-binding proteins in their systems to rid them of the deadly metals.

Cold seeps are areas where gases seep through the rock in Earth's crust and emerge on the ocean floor. Millions of tubeworms, cousins of the tubeworms living near hot vents, live near cold seeps. Unlike hot vent tubeworms, which have short life spans, cold seep tubeworms can live up to 250 years. Except for giant clams, they may live longer than any other invertebrate. Bacteria thrive around cold seeps. There, as near hot vents, they are the foundation of marine communities.

Cold seep creatures include ice worms, squat lobsters, and snails. Extra large mussels have also been found in cold seeps. Mussels push themselves around with one foot. They attach themselves to rocks and to each other with threads. The mussels stay still for long periods of time, waiting for bacteria to pass over their gills. They do not eat the bacteria. They, like tubeworms, depend on the sugars bacteria produce for food. If the cold seeps stopped releasing the gases that allow the bacteria to survive, the mussels would be doomed.

Bacteria get energy from gases released at cold seeps.

Mussels around cold seeps have large gills where hundreds of sugar-producing bacteria live.

Other Sea Creatures Adapt and Survive

GULPER EELS have huge mouths and elastic cheeks that allow them to swallow creatures three times their size. A gulper's jaws make up 80 percent of its weight. Gulpers also have a luminous tip on their long, whiplike tails that shines with a reddish glow. Researchers think gulpers may lure their prey with these glowing tails. Gulpers live 6,500 feet (1,980 m) below the surface.

ELEPHANT SEALS dive 1 mile (1.6 km) deep to catch ratfish, their favorite food. Elephant seals and sperm whales are among the few mammals known to dive that deep. The seals can do this because of their streamlined shape and ability to conserve energy underwater.

HAGFISH are scavengers. They eat the carcasses of large fish and then coat the leftovers with a disgusting slime so other scavengers won't eat them. They can live as deep as 3,000 feet (914 m) underwater.

VAMPIRE SQUID live their entire lives up to 3,000 feet (914 m) underwater. Webbing between their tentacles makes these squid look as if they are wearing capes. Together with their skin, which can appear jet-black, and their eyes, which can appear deep red, they really do resemble the vampires we see in movies. Vampire squid can escape predators by ejecting a cloud of mucus filled with a thousand tiny spheres of blue light. The light spectacle confuses predators, giving the squid an opportunity to disappear into the darkness.

TRIPOD FISH stand on the ocean bottom on three fins that look like stilts. These fins let the fish move on the ocean bottom without sinking into the soft ooze. Tripod fish can live as deep as 18,000 feet (5,486 m) underwater.

BRITTLE STARS have arms that often break off when a predator attacks, but new arms grow back quickly. Brittle stars' feet are just as remarkable. They help brittle stars breathe, put food in their mouths, and keep a firm grip on the ocean floor. Brittle stars can be found from the shore down to 6,755 feet (2,059 m).

GLOSSARY

BACTERIA — organisms made up of one cell. They are among the smallest life forms.

CAMOUFLAGE — coloring or body shape that makes an animal hard to see in its natural surroundings

COLD SEEP — where gases seep through Earth's crust and emerge on the ocean floor

DORSAL — near the back

ECHOLOCATION — the use of echoes to locate objects

ENVIRONMENT — surroundings of a person, animal, or plant that affect its growth and development

GILLS — organs that take oxygen from water as it passes over them, enabling certain kinds of sea creatures to breathe

HABITAT — the environment in which an organism usually lives

HOT VENT — where a stream of mineral-rich hot water gushes out of an opening in the ocean floor

INVERTEBRATE — an animal without a backbone, such as a jellyfish

LUMINOUS — bright, or giving off light

OXYGEN — a gas found in air and water that most animals need to survive

PHOTOSYNTHESIS — the process by which plants produce food using energy from sunlight. Oxygen is a by-product.

PREDATOR — an animal that hunts other animals for food

PREY — an animal that is hunted by other animals

SCAVENGER — an animal that feeds on decaying meat or garbage

FIND OUT MORE

Books

Aquatic Life of the World. Tarrytown, NY: Marshall Cavendish, 2001.

Grupper, Jonathan. *Destination: Deep Sea*. Washington, DC: National Geographic Society, 2000.

Kinchen, James. *Squids*. Danbury, CT: Grolier Educational, 1999.

Lambert, David. *The Children's Animal Atlas*. Brookfield, CT: Millbrook Press, 1992.

Martin, James. *Tentacles: The Amazing World of Octopus, Squid, and Their Relatives*. New York: Crown Publishers, 1993.

Pope, Joyce. *Life in the Dark*. Austin, TX: Steck-Vaughn, 1992.

Websites

Extreme Science Adventure: Studying the Ocean Deep
http://www.extremescience.com/howdeep.htm

Monterey Bay Aquarium Website: Frequently Asked Questions about the Deep Sea
http://www.mbayaq.org/efc/efc_se/se_exp_faq.asp

NOVA Online: Into the Abyss
http://www.pbs.org/wgbh/nova/abyss/

Smithsonian's National Museum of Natural History: In Search of Giant Squid
http://seawifs.gsfc.nasa.gov/OCEAN_PLANET/HTML/squid_opening.html

University of Washington Exploraquarium: Deep-Sea Hydrothermal Vents
http://www.ocean.washington.edu/people/grads/scottv/exploraquarium/vent/intro.htm

Woods Hole Oceanographic Institute: Dive and Discover
http://www.divediscover.whoi.edu

INDEX

Page numbers in **boldface** are illustrations.

adaptations, 10, 17, 26–27, 30, 31, 32, 33, 34, 35
anglerfish, **8–9**, 15, **15**

bacteria, 10, 18–19, 26, 28, **28**
bioluminescence, 12–13, 14, 15, 17–18, 30, 33
black dragonfish, 13
black smokers, 26
bottom dwellers, **8–9**, 22–23
brittle stars, 35, **35**

camouflage, 14, 18
clams, 23, 26, 28
cold seeps, 27–28, **28**, **29**
communication, **21**
crabs, 5, **5**, **8**, 26

defense mechanisms, 12, 14, 18, 32, 33
dolphins, 5

echolocation, 19
eels, **9**, 30, **30**
eyes, 17, 19, 23

flashlight fish, **10**, 12
flatfish, 23, **23**
food, 10, 12–13, 14, 15, 17–18, 19, **22**, 23, 26, 28, **28**, 30, 31
food chain, 26

gases, 5

hagfish, 32, **32**
hatchet fish, **9**, 14, **14**
hot vents, **24–25**, 25–27
hypothermal vents, **4**

jellyfish, 5

lifespans, 28
light. See bioluminescence; sunlight zone; twilight zone
lobsters, 28
locomotion, 17, 22–23, 28, 34, 35

marine snow, 10, **22**
metals, 5, 26–27
minerals, 5, 25–26

mussels, 28, **29**

oxygen, 7

parenting, 18–19
photosynthesis, 5–7
plankton, 10
plants, 5–7, **6–7**, 10, 26
poisons, 5, 26–27
predators. See food
pressure, 5, 10, 23
prey. See food

rattails, **9**, 23, **27**
recognition, 14
regeneration, 35
reproduction, 15

scavengers, 32
seals, 30, **30**
sea pens, **22**, 22–23
sea stars, 5
sharks, 5, **8**
shore, 5, **5**
shrimp, 12, **13**
snails, 28
sound, 19
sponges, 23
squid, **8–9**, 10, **12**, **16**, **17**, 17–20, **21**
    giant, **18**, **19**, 19–20
    vampire, 33, **33**
starfish, 23
    See also brittle stars; sea stars
submersibles, 10, **11**
sunlight zone, 5–7, **6–7**, 26

trenches, 10
tripod fish, 34, **34**
tubeworms, **9**, 26, **26**, 27–28
twilight zone, 7–10

Websites, 38
whales, 5, **9**, 19
worms, 23, 28
    See also tubeworms

ABOUT THE AUTHOR

Wendy Pfeffer, an award-winning author of fiction and nonfiction books, enjoyed an early career as a first grade teacher. Now a full-time writer, she visits schools, where she makes presentations and conducts writing workshops. She lives in Pennington, New Jersey, with her husband, Tom.

For Mom and Dad Pfeffer, who loved the ocean.

With special thanks to Kate Nunn, for gently pushing me to write my best.

Thanks to Paul Sieswerda, for his expert review of this manuscript.

Benchmark Books
Marshall Cavendish
99 White Plains Road
Tarrytown, New York 10591-9001

www.marshallcavendish.com

Text copyright © 2003 Wendy Pfeffer
Illustrations by Sonia Chaghatzbanian
Illustrations copyright © 2003 Marshall Cavendish Corporation

All rights reserved. No part of this book may be reproduced or utilized in any form or by any means electronic or mechanical including photocopying, recording, or by any information storage and retrieval system, without permission from the copyright holders.

Pfeffer, Wendy, 1929-
Deep oceans / by Wendy Pfeffer.
p. cm. — (Living on the edge)
Summary: Examines the harsh living conditions that exist deep below the surface of the ocean and describes the animals, including the squid, hatchet fish, and tubeworm, that have adapted to those conditions and have made the deep parts of the ocean their home.
Includes bibliographical references and index.
ISBN 0-7614-1439-8
1. Deep-sea animals--Juvenile literature. [1. Marine animals.] I. Title. II. Living on the edge (New York, N.Y.)
QL125.5 .P44 2003
591.77--dc21
2001008754

The photographs in this book are used by permission and through the courtesy of: *Animals Animals*: E.R. Degginger, title page, 26; Bob Cranston, 16, 21, 31; Kuiter R. OSF, 22. *Photo Researchers*: B. Murton/Southampton Oceanographic Centre/Science Photo Library, 4; Tom McHugh, 32; Robert Dunne, 35; P.M. David, back cover. *Corbis*: Farrell Grehan, 2; Robert Yin, 6–7, 23; Bruce Robison, 12 (bottom), 14; Stephen Frink, 17; *Reuters NewMedia Inc.*, 19; Ralph White, 24–25. *Peter Arnold, Inc.*: Norbert Wu, 10, 15, 30, 34. *Harbor Branch Oceanographic Institution*: 11, 12 (top), 25, 33. *Woods Hole Oceanographic Institution*: Holger Jannasch, 28; Carl Wirsen, 29.

Photo Research by Candlepants, Inc.
Cover Photo: Peter Arnold, Inc./Norbert Wu
Book design by Sonia Chaghatzbanian
Printed in Hong Kong

1 3 5 6 4 2

ML